MAKING AFRICA WORK THROUGH
THE POWER OF INNOVATIVE VOLUNTEERISM

DR. RICHARD MUNANG

authorHOUSE®

AuthorHouse™ UK
1663 Liberty Drive
Bloomington, IN 47403 USA
www.authorhouse.co.uk
Phone: 0800.197.4150

Published by AuthorHouse 05/03/2018

ISBN: 978-1-5462-9240-1 (sc)
ISBN: 978-1-5462-9241-8 (hc)
ISBN: 978-1-5462-9239-5 (e)

Chapter 1

THE BEGINNING

My African Roots—Africa Born in Me

The year is 1978. It was a glorious time for my little village, Jinkfuin, in the north-western part of my country, Cameroon. I was born in October, a time of harvesting. I am the second to the last born in a family of five siblings. Life in Jinkfuin typified the common traditional African setting. There was a chief who ruled in union with a council of elders. Everyone was helpful to each other and hospitable to visitors to the village, despite the injection of Western education systems. A child belonged to the community, and the whole community participated in raising that child. My arrival into this world was characterised by the usual celebration of joyous revelling and feasting.

1

The people in my village were small-scale farmers, and cassava, maize, and bananas were their main crops. Our village epitomised the African values of togetherness and loyalty to the various cultural festivals we, as a people, performed to mark different occasions. These traditions were quite vibrant, as they were passed down from generation to generation. It was the duty of the village elders to teach the upcoming generations. Growing up, I looked forward to these teachings, where I got to learn the history of our people and the values we hold dear. When I was quite young, about six years of age, others my age and I gathered around a fire in the night as one of the elders told us stories. The stories were tailored to teach us important values to aid us in our later lives. Values such as respect of elders, hard work, diligence, humility, integrity, and unity were imparted to us at an early age through these nightly stories.

When I was seven years old, I was enrolled in the local primary school, Jinkfuin Government School. It was several kilometres from our village, so it required that I wake up quite early to walk to school. To complete primary school, one needed to complete seven years. So I hurried to school every day for seven years without shoes. I was at first very reluctant to go to school because some my age were not enrolled in the school. I envied them because they spent most of their day having all sorts of fun. As the years progressed, though, I began to appreciate going to school and learning new things. Learning to read and write was of great value to me. I particularly took a liking to mathematics.

It was while at school that I struck a friendship with a quiet boy called Nsom. We became best of friends and were inseparable, doing most things together. Nsom hailed from the next village that was on my way to the school. I passed by every morning so we could walk the remainder of the way to school together. During holiday breaks, I joined the other boys of our village to herd our families' goats. We set out in the morning, each armed with his packed lunch, and headed deep into the forest, leading the animals before us.

The years went by, and as many before me, my time came to be initiated to young adulthood according to our traditions. This was an important transition in the lives of many of us in the village. It was marked with great pomp and feasting. Young men including myself, who were to pass over from being children to young adults were arrayed in the traditional garb and spent the eve of the day of celebration secluded from their families. It was at this point that we got to meet the chief and the council of elders. They shared with us the significance of the initiation and the responsibilities we would assume in our new status. The following morning, we were paraded before the whole village. After the ceremony, the whole village broke into song and dance. Later in the afternoon, we sat to eat together. Here, I learned to like my people. They rejoiced with us for our achievement.

As we danced, I felt a sense of pride and accomplishment – having crossed over to adulthood. We spent our time engaged in traditional wrestling games. To win, one had to form the strongest alliance; I was quite good at forming these alliances.

These games went on till evening, when each of us retreated for supper. We regrouped later in the night for a session of storytelling with the elders. We all huddled together around a fire and told stories till late in the night.

Back in school, it was the final year of my primary education. This was a crucial time because a national examination at the end of the school year would determine whether I would advance to the secondary level. I was particularly keen to advance because I was eager to learn more. I was hungry for knowledge, so that year I spent most of any free time I had studying. Nsom and I always sat under one particular tree during the lunch break to study together. The exams would soon be upon us, and it was a three-day event. When the exams were over, I was very optimistic that I had aced them.

The results of the national exam took a month and a half to be released from the national examination council. I spent the time contemplating how my life would be, since I was confident I had done well. As I came in from herding the family goats one evening, I was surprised to see my schoolteacher at our home. I knew straight away she had come bearing news of my exam results. She broke the news that I had passed quite well, being amongst the best students in our primary school. With these results, I was able to gain admission into Baptist Comprehensive College, a secondary school. I studied there for five years, as the tradition of the curriculum circle entailed, and graduated with ten ordinary-level papers in the sciences, the best in the school that year. My parents were very happy and proud of me.

I was overjoyed and thanked God my dream of joining a high school was coming true. But my parents didn't have the money to send me to high school. My father's brother, fondly known as Pa Tingem, intervened and got me admission to Government Bilingual High School (GBHS) in Bamenda. The next day, I visited Nsom to share my good news and learn how he performed. Nsom had also done well, though he was not called to the same school. We spent that afternoon daydreaming how our lives would be in our respective schools. We shared the hopes we had for the future regarding the careers we would like to pursue. Evening soon came. I returned to my village, feeling happy and contented.

The year was coming to an end. For me, the new year held the prospect of leaving the village and heading to the city of Bamenda. I was beside myself, waiting for the day of my departure. I reflected on my early childhood, the values I learned, and the rich African heritage imparted by my elders. I smiled, thankful I was born in this part of the world. Africa was truly born in me during these impressionable years.

Chapter 2

HIGH SCHOOL AND UNIVERSITY EDUCATION: FROM VILLAGE TO CITY

Behold, the Elders Left the Right Door Open

A quintessential African proverb best captures my transition to the city: "The worlds of the elders do not lock all the doors; they leave the right door open." While I enjoyed village life, I was oblivious of the dim prospects that my continued stay in the village presented at that time. Yet the one who created me knew how to lead me out of it and presented me with opportunities to ensure such limited prospects never confronted another child—another village girl or boy. Future village girls or boys would not be forced by circumstances to abandon the beloved town of their births in search of opportunities; they would be confronted by opportunities

7

right on their doorsteps. The choice to live in or depart the village could be made at their pleasure.

The right door opened for me to get a quality education in the bustling city of Bamenda, and there was no looking back. The day was finally here. My long journey to the city was at hand so I could continue with my education. I was very excited and nervous at the same time. Here I was, a budding teenager from the village, venturing to the big city of Bamenda. I had heard stories of how life was quite different in the city. This was the cause of my nervousness. But I was excited at the prospect of continuing my education. I could not wait to enrol at GBHS. I was accompanied by my favourite uncle, Pa Tingem. He had travelled from the city to the village because of how well I had passed my secondary level. We boarded the bus, and the journey began as I bade goodbye to my family and Nsom, who was also preparing to travel to his high school.

The long journey was tiring, but we finally arrived in Bamenda. I was immediately struck by the different landscape of high-rise buildings, paved streets, and heavy vehicle traffic on Bamenda's commercial avenues. It was around sunset, and the streets were packed with vehicles and people heading home after a hard day at work. We went to my uncle's place, where I stayed for my two years in GBHS and while at university. The next morning, we reported to my new high school. I couldn't wait for morning; I hardly slept that night, imagining how my school would be.

Morning finally came, and my uncle took me to GBHS Bamenda. Once enrolled, the situation at school was tough, due to the few resources at my disposal. I lacked shoes. I thought I could borrow from my elder brothers, but they would have none of it. Whenever home, in the event I got the opportunity to use their shoes and come back with them muddy, I would later see red because of the beatings I received from them. Now looking back, though tough, it was their way of teaching me to stay focused on what was important. Times were hard. I remember once when I had to trek part of the journey to school from the holidays—a total of seventy kilometres—due to lack of fare.

All this was bearable to me due to the hunger I had for learning. I had taken to books as a duck takes to water. I was so glad to be in what I regarded as the best high school in the country. The tutors I had took an immediate liking to me because of the zeal I had to learn. This taught me an important life lesson in that it takes the support and input of others for one to succeed at any level. I constantly asked my tutors questions which caused many of them to really go deeper in the things they were teaching. It is true that when you truly want something, you will persevere to the end to finally have it. Despite all the hardships, I endured to the very end to pass with the highest grade.

My joy for passing my GCE Advanced Level was cut short when it became apparent that the fees to further my education to the university were lacking. Relief came when Uncle Pa Tingem, having seen how well I had done at the secondary and high school level, volunteered to invest in my education. He saw the promising return the family would have in financially supporting me through my university education. I got admitted to the University of Yaoundé I Higher Teachers Training College, fondly known as ENS Bambili, to pursue a bachelor's degree in physics and education. I lived with my uncle's friend for the three years I was at the university because I lacked the finances to rent my own room. I trekked every day seventy kilometres to the university for the three years I was there. This was at the turn of the millennium, in the year 2000.

Life in the university was invigorating. Socially, intellectually, and educationally, it brought to me a very different outlook to life. Socially, I witnessed the effect of Westernisation in the hostel corridors as young men and women endeavoured to ape the Western culture in dressing and public behaviour. I was grateful for my traditional African upbringing; it kept me objective in what to borrow and not to borrow from the Western culture. It kept me focused on what was important: my books. Intellectually and scholastically, I was intrigued by the depth and vast knowledge that my young mind could absorb. I spent most of my free time in the library, studying and perusing the great works of those who had gone ahead of me. I kept well away from the groups that engaged in revelling, hopping from one party to another and drinking all night, behaving uncouthly.

The end of the first semester saw me travel back to the village to see the folks back home and herd the family goats. There was no one around to perform this task, and it fell on me, being the youngest son in the family. This is the time when I saw the devastating effects the erratic weather—climate change—was having on farming in our village. It really frustrated me to witness my mother's tireless efforts in tilling our farmland, only met by failing rains and much diminished harvest. To see old rivers ebb away. It got me curious, why was this happening and what could be done to salvage the situation. Many questions arose in my mind, and it tweaked my interest to delve deeper into the climate and environment phenomena. It proved very tough, balancing studying and herding the family goats, but by the grace of God, I endured, knowing a far greater reward awaited me soon.

I returned to the capital to resume the second semester of my degree programme. Something was birthed in me regarding the issue we now know as the changing climate and its effects on the livelihoods of ordinary people. The second semester breezed by quickly, and the end-year exams were soon upon us. We quickly embarked into the second year of the teachers training programme, without breaking for holiday. Then the third and final year; my diligence, fuelled by the responsibility I bestowed on myself to ensure the problem of diminishing yields, was solved. The questions about the climate constantly occupied my mind. I determined to pass my degree course with highest distinction so that I may be able to pursue a master's degree in environmental science. It was going to be a tough order, but there was a fire burning in me and a strong belief that I could make it.

The final exams came, and after sitting them, I was confident that I had passed. I awaited the results eagerly, knowing deep inside I had outdone myself. Indeed, to no surprise, I emerged amongst the top of my class in physics and education. I was in seventh heaven; my diligence had paid off in a great way. There was great celebration back home for my great achievement. After the celebrations and the euphoria had died down, I quickly embarked on applying for scholarships as I awaited to be posted for a teaching job in a secondary school around the country. Months went by with no positive response in my numerous scholarship applications, and no job posting. It was a trying time for me, in that some of my fellow graduates were already working, while nothing good was coming my way. After a year, I decided to volunteer in a local non-governmental organisation (NGO). The time spent working with the NGO served me well, as it enabled me to learn vital practical skills in project development and management.

A colleague I was working with informed me about a competition for least development countries which was ongoing in Nottingham University, where the winner could get a scholarship to the university. Finally, this was the opportunity I was longing for. I was determined to come out victorious. Dreams do come true when you have focus and work diligently towards the passion burning within you. Using the experience I got from volunteering with the NGO, I framed an application and submitted it. Months passed by, and finally, I emerged the winner of the competition; my dream to pursue a master's degree in environmental science was in my grasp. I was beside myself with the joy I felt at having a chance to finally seek answers to the questions that had plagued me after seeing the reduced harvest and crop failure back at the village.

Chapter 3

FROM CAMEROON TO ENGLAND

Choices Have Consequences; Dedication Pays

Once again, my father, brothers, sisters, and nephews escorted me to the Douala Airport to board a plane to England. This was my first time to ever be in Douala. I was looking forward to continuing with my studies after a whole year since I graduated from ENS Bambili. I was privileged for the time I volunteered at the NGO, where I gained valuable working experience, refined my technical skills, and rubbed shoulders with a great many individuals to further sharpen my interpersonal skills. I really appreciate my colleague Mr Julius Nde, who informed me of the opportunity to frame a good scholarship application to win a scholarship to study in Europe. True to a Cameroonian proverb, "He who asks questions cannot avoid answers," I was diligent in my work and was very inquisitive to learn and to gain more knowledge in the field of environment science. I took the opportunity, and it bore the fruits of me getting the opportunity to further my studies. So here I was; an opportunity to have my questions answered had materialised.

At the lobby of the airport, I said goodbye to my uncle, mother and dad, sisters and brothers, and the entire family and some of my colleagues from the NGO and walked towards the security desk. I was quickly cleared and ushered through to head to the plane. I was very excited, being the first person in my village to be flying abroad. It was indeed an historic moment for the people of Jinkfuin for a son of theirs flying out to Europe—knowing fully well that it was not in vain; something good would come out of it. I was welcomed into the plane by a beautiful air hostess and ushered to my seat. I was in luck in that I was seated on a window seat and watched as the plane taxied on the runway, picked up speed, and lifted off. It being my first time flying, I felt awkward during the take-off, but on reaching the cruising attitude, I started enjoying the experience. Touching down at London's Heathrow Airport, I again felt awkward and felt goosebumps all over my body. Here I was, arriving in a foreign land for the first time in my life. In exiting the plane, the first thing that hit me was the cold air.

I was glad that I was met at the airport by someone from the university, who introduced himself as Mr O'Connor. He courteously inquired whether the journey was okay as he welcomed me to his country. Unfortunately, my luggage didn't arrive, and I was tense, given that I didn't know what to do and whether it would ever arrive. Mr. O'Conner explained the situation, and I managed to stay calm. We then drove straight from London Heathrow to the University Park Campus, the University of Nottingham residence, so that I may rest awaiting the following day, when Mr O'Connor would take me on a tour of the campus, coupled with registering for the course I planned to do. I was quite impressed with the planning, preparations, and hospitality they accorded me. Bearing with the cold weather, my first challenge came when he took me to the college hall to have lunch. The variety of foods offered was not what I was used to from back home, but I settled for a plate of rice with beef stew. I was glad by the help of God I was quickly able to adjust to the weather and diet being offered.

The lectures began in earnest, and it was like a breath of fresh air engaging in the things that had made me travel all the way from Cameroon. The discussions we shared amongst ourselves and the private discussions I had with the various professors was eye-opening and intriguing, feeding the deep hunger I had for environmental science. The research work was tedious but bearable because of the passion burning in me.

There came a time we had a lecture from Professor Jeremy Colls, who took us through a session in our course study. I took an immediate liking to his mastery of the modelling lesson he was taking us, from the way he made modelling very relatable to the real issues we faced back in my village. I knew full well I had to work with him. Later, I approached him, and we had a lengthy discussion with him regarding crop failure in Africa. I shared my experience of what I had seen back home, and he enlightened me as to the various possible causes of the reduced crop harvest and crop failure in Africa and around the world. This triggered my interest in modelling agro-ecosystems, which became my focus of study.

I poured myself wholly towards studying all that I could find and researching agro-ecosystems. My diligence once again paid off, in that I graduated from the master's programme with distinction. The heavens again smiled upon me, as I won an overseas research scholarship for a PhD programme, which was to run for three years and was soon to begin. Unfortunately, I had no chance to travel to Cameroon to celebrate with my people. This minor disappointment was overshadowed by the burning passion within me, which daily convinced me without a shadow of doubt that a solution would be forthcoming concerning the problem of global climate change and its impact.

The PhD programme meant I had to seek outside residence from the university, which necessitated me to seek for a place to rent. This placed a strain on my finances, and I had to seek employment to pay for rent and my personal upkeep. The job-seeking task was a bit frustrating, in view of my limited work experience. I managed to get a job in a mail centre called Royal Mail, where I sorted out mail. The work was tedious, but it made it possible for me to have a roof over my head, food on my plate, and clothes on my back.

I was progressing well in my PhD programme, as I gained more and more insight and revelation into the issues concerning climate change and agro-ecosystems. My efforts paid off, and I did it in two years instead of the initial three. I was a PhD after two years and was given an opportunity to lecture at Nottingham University. Later, I went to Trinity College Dublin as a senior researcher. This was a welcome relief for me because it meant I did not need to continue work at the Royal Mail. It also provided me with the opportunity to engage in my passion and interact with young upcoming minds in the field of global climate change and its impact on the environment. True to the African proverb, it takes a whole village to raise a child. It would indeed take the effort of all of us to come up with viable solutions to solve the climate change issue, since it affects us all.

This got me thinking that besides my input in nurturing upcoming scholars in the field of environmental science, I needed to reach out to the world. This elicited my quest in seeking a job to use my knowledge and skills. I therefore applied for a vacancy in the UN Environment Programme (UNEP). I was called for the interviews, and after going through a series of interviews, passing each one, I finally landed the position. I was ecstatic; I now had a chance to impart my generation, and I gathered a few of the friends I had made to go celebrate the good news.

More good news followed when I was posted back to Africa in Nairobi, Kenya. I spent the few days before my posting to Kenya saying my goodbyes and thanks to the fraternity at the University of Nottingham. I was quite surprised when they threw me a farewell party and showered me with gifts and mementos, on top of speaking very well of me. It was refreshing.

Chapter 4

AFRICAN EMBODIMENT; REALISING AFRICA'S DEVELOPMENT

"Even in extreme drought, the lion does not succumb to eating grass because that's against its nature."

I recently read the above quote on Twitter—no doubt by one of the many ambitious young people Africa is blessed with. It reminded me of my own life. Facing difficulties but confronting them boldly and succeeding. This is the resilience of the African people demonstrable in many ways.

I therefore wish to make a quote myself: "A rising Africa cannot succumb to purported challenges. That's against its nature."

For far too long, Africa has been thought of as the "Dark Continent". This is very erroneous in that light has broken forth through the efforts of individuals who have graced the global stage and made such great impact in world affairs, like the late former president of South Africa Nelson Mandela, former UN Secretary General Koffi Annan, Noble Laureates Wangari Mathai and Desmond Tutu, and the late Thomas Sankara of Burkina Faso, who under difficult circumstances of an ailing economy, a very low economic base to start with and minimal to no international support, raised literacy levels of his country from 13 to 73 per cent in just four years and raised wheat production by 120 per cent, making the country food secure in three years. The resilience and drive of the African people quite amazes me and fascinates me. These people have endured the brunt of slavery and colonisation. In North America, the likes of Martin Luther King, Rosa Parks, Malcolm X, Muhammad Ali, Michael Jordan, Tiger Woods, and Barack Obama have risen high above their peers in their spheres of influence. The recent story of former football star George Weah, now president of Liberia, speaks volumes. Truly, the difference between dreams and reality is having a plan of action and executing that plan.

Time and opportunity happen to all of us; what makes the difference is execution of a plan of action. True to the Cameroonian proverb, "Rain does not fall on one roof alone." The thing that distinguishes is how one uses the rain. The ingenuity and potential in an individual only represents opportunity, which is not enough to achieve transformation and if not harnessed will remain a missed opportunity. This then means that opportunity must be harnessed to realise value.

To this end, I have taken it upon myself to ensure Africa's potential, residing in its people, can be realised for the benefit of many. My experiences have taught me that an empowered people across the continent is the best bet towards accelerating socio-economic development and warding off vulnerabilities.

Agriculture, which is highly vulnerable to climate change, continues to be Africa's primary socio-economic sector. I have taken a keen interest to combining these two areas, as a chemist would a concoction, but this time towards premising agriculture as the engine of socio-economic transformation in Africa and ensuring the continent's people lead this pursuit. I have considered the combination of these areas with energy and enhanced policy and operational aspects: technical, tactical, and technological capacities for countries to implement much-needed climate actions in an integrated manner, where outcomes maximise both socio-economic development priorities and the need for building climate resilience.

My work as a professional in climate change has focused on making it relatable to the ordinary person, through premising climate action as a solutions provider. I have championed a strategic approach to implementing climate actions across Africa that prioritises catalytic sectors of agriculture and clean energy capable of meeting leading socio-economic priorities of food and livelihood security simultaneously with enhancing ecosystems and offsetting carbon to meet climate objectives across the continent.

Knowing full well that people are the greatest resource we have in Africa, I have prioritised in my work, setting structure to harnessing Africa's human capital in its diversity towards accelerating socio-economic transformation and climate resilience. To tap into the resilience of the African mind and heart as the source of wealth for the continent.

As seconds turn to minutes, minutes to hours, hours to days, days to months, and months to years, all people of good will towards Africa need to ask some tough questions. Each of us needs to assess what Africa has achieved to date and what could be done differently to make a better outcome. For instance, it is documented that an excess of $15 billion has been invested in Africa's agriculture over the past two decades. Yet we still grapple with the same challenges we did twenty years ago. Africa still cannot feed itself.

Cumulatively, there are a total of fifty-eight major energy initiatives in the continent. Through such initiatives, a total of $30 billion has been invested from multilateral and bilateral sources in just four years (this is excluding the private sector). Yet over 60 per cent of Africa is not productively energised.

One starts to ask whether money is really the biggest problem in the continent, as it is made to seem.

As we transition to the meat of this book, as an out-of-the-box thinking climate scientist, I have asked myself one big question: How can climate change be the silver bullet, the master key that unlocks the door to accelerated socio-economic transformation in the continent? How can it be relatable to the development economist, planner, thinker, and commoner on the street?

The tragic flood in Sierra Leone in August 2017 devastated me. I had only recently been to the country, so the distress shook me to my very core. I cannot help but wonder at the level of devastation of ordinary Sierra Leoneans. I feel for them. This tragedy got me thinking, and as unfortunate as it was, it provided us an opportunity to attempt a summarised answer, a case study to decouple several issues into a coherent solution that combines climate action and accelerated socio-economic transformation.

Making African Resilience Count: Turning the Sierra Leone Flooding Crisis into a Lasting Solution

Let's put things into perspective. In Sierra Leone and across Africa, the science is unequivocal: Climate change is a contributory factor, alongside human-made elements like deforestation and encroachment, to the kind of disaster that hit Sierra Leone in August 2017. This is no longer an abstract issue.

According to the US National Weather Service's Climate Prediction Center, Sierra Leone received an unprecedented amount of rainfall in 2017: three times the normal seasonal rainfall. Such torrential rains are a clear sign of the changing climate. In August 2017, at the height of the rainy season, Freetown received an unprecedented average of 539.9 millimetres of rainfall. With its land size of 356.9 square kilometres, Freetown had an average of 190 million cubic metres of rain water to drain. This extreme volume of water, combined with human factors like encroachment on natural environment such as creeks and wetlands, which are the natural drainage and storage systems for flood waters, as well as construction on flood-prone areas and inefficient drainage systems, led to this disaster.

Scientists say that no city in Africa is immune to extreme climate change effects.

In early 2017, new data from the UK Met Office, the United Kingdom's national weather service, and the US National Aeronautics and Space Administration showed that the earth's temperature had increased to about 1.1 degrees C above pre-industrial levels. This is dangerously close to the 1.5 degrees C threshold set by the Paris Climate Change Agreement to prevent the worsening effects of climate change. At this rate, Africa's coastal cities are vulnerable. Sea level rise is projected to hit coastal cities—14 per cent higher than the global average by 2100 for the fast approaching over 4 degrees C warming scenario. The impact will stretch far and wide beyond Freetown to expose millions to risk of flooding. By 2050, high numbers are projected in coastal cities of Mozambique (five million), Tanzania (two million), Cameroon (two million), Egypt (one million), Senegal (0.5 million), and Morocco (0.5 million). Such flooding will reverse economic and development gains with the ensuing health impacts and damage to infrastructure, loss of tourist sites, and disruption in food supply. It will also expose the populations to elevated food prices, loss of livelihoods, and strife.

As far as ecosystem degradation is concerned, Africa loses up to $68 billion annually. This means that the continent's natural buffer against such impending climate change effects is being lost at a rate of about $180 million daily.

The science is clear. The escalating climate change knows no boundaries, and this is coupled with an increasingly degraded environment. Countries across Africa need to urgently address these dual challenges if we want to forestall similar disasters in future.

This is the logic behind the universal, global response to climate change that has been called for under the Paris Agreement. Sierra Leone, classified as the third most vulnerable country to climate change, stands to benefit from being part of this global collective action, and it is among the countries that have ratified the Paris Agreement, demonstrating its resolve to combat climate change. In fact, Africa has shown global leadership in responding to climate change.

Combining Ecosystem Restoration with Alternative Livelihoods

The good news is that practical solutions have been successfully applied across Africa. In Rwanda's Geshwati area, a land suitability map informs policy decisions to relocate vulnerable communities from previously encroached natural environments and high-risk areas to safer habitation areas. Considering that agriculture is the backbone of these communities, this plan also takes advantage of ecosystem-based adaptation (EBA) agriculture techniques that the communities can safely engage in for their livelihoods. Simultaneously, the plan is guiding the restoration of previously degraded catchment areas using EBA techniques like planting indigenous trees to stabilise soils and slopes and to regulate floodwaters. This has eradicated landslides that were once a common phenomenon in the area.

A similar two-pronged strategy has been successfully applied to build resilience in Mozambique's coastal communities that were highly vulnerable to coastal flooding. For example, an investment of $120 per person to rehabilitate depleted mangroves and establish crab farming. These have become natural buffers against coastal flooding while simultaneously preventing future encroachment by providing alternative livelihood activities away from the mangroves.

Sierra Leone and other at-risk countries can benefit from similar strategies that restore degraded ecosystems, allowing them to act as a buffer to the compounding effects of climate change. Creating sustainable alternative livelihood activities away from risk-prone areas can also help prevent possible future encroachment, thus preventing the degradation of natural ecosystems. After all, natural ecosystems are our best bet against mounting climate change-driven disasters.

We must also fight the trend of urbanisation.

Africa's cities face the fastest pace of population growth globally. This growth, however, does not reflect positively on economic growth, which is a key enabler to building climate resilience. For example, the World Bank notes that African cities are almost 30 per cent more expensive than other countries at similar income levels. Housing is 55 per cent costlier, and food prices are 35 per cent higher than in other countries. Considering the high unemployment and underemployment, more than 50 per cent of urban dwellers end up living in slums. Sierra Leone, which faces an urbanisation rate of 2.9 per cent, has 75.6 per cent of its urban population in informal settlements. These urban poor stand out as the most vulnerable, something that needs to be urgently addressed.

To address this, a key area is diversifying and decentralising socio-economic growth opportunities away from cities. This is critical to eradicate the allure of cities as being the only areas with income opportunities. It is vital to decongest cities and curtail the growth of informal settlements that are vulnerable hot spots.

Focusing on ecosystem-based adaptation agriculture and industrialisation powered by clean energy offers an opportunity to diversify income opportunities into sustainable sectors like clean energy. Cumulatively, this amalgamation is projected to create an agro-industrial sector worth up to $1 trillion by 2030, while ensuring ecosystems are taken care of and carbon is offset to ensure climate resilience. It is such diversification that will open rural Africa, where 70 per cent of agriculture takes place, to industrialisation and the creation of economic opportunities to relieve the pressure on urban areas.

How to Make It Happen

Making this paradigm shift happen requires a collective undertaking. It will take the intervention of both state and nonstate actors, as is called for in the Paris Agreement. Mutual partnerships will have to be formed to bridge policy and operational gaps. Through the Ecosystem-Based Adaptation for Food Security Assembly (EBAFOSA), a pan-African initiative that you will read more of, countries and stakeholders are engaging to bridge critical gaps. Among the intervention areas that are being prioritised is the harmonisation of policy across complementary ministries, such as the ministries of agriculture, industrialisation, land, energy, trade, and roads, among others, to establish clean energy-powered agro-industrial zones to create sustainable jobs.

For example, countries are harmonising finance, industry, energy, and agriculture policies to establish tax incentives for agro-based industries that are powered by clean energy in rural areas. These are set to attract investment to these areas to fuel job creation and take the pressure off urban centres like Freetown.

Another area is affordable financing. To fuel the growth of sustainable businesses that promote climate resilience, EBAFOSA is facilitating partnerships between financers and actors to develop risk-sharing facilities targeted at de-risking lending to the catalytic area. These risk-sharing facilities provide money to cover default risks and unlock up to ten times the securitised amounts for lending to entrepreneurs supporting sustainable clean energy-powered agriculture. This risk-sharing facility is reducing the cost of capital and incentivising private sector actors engaged along this chain— from farmers, distributors, marketers, and advisory service providers—to capitalise their businesses and create sustainable jobs to lure people away from crowded cities.

Another important area where harmonisation can happen is in the building of complementary partnerships with multiple stakeholders to bridge policy and operational gaps towards sustainable rural industries. In rural Cameroon, EBAFOSA is catalysing partnerships linking off-grid small-hydro systems to power cassava and Irish potato processing. These are then linked to markets and supply chains across the country, using mobile apps. A total of ten youth groups working in information and communications technology (ICT), clean energy, and marketing have been engaged, creating green jobs for approximately a hundred young people. More than five hundred women now have access to value-addition services. As a result, they have cut their post-harvest losses (PHLs), enhancing their income stability and increasing the community's food security. Such gainfully engaged people will not be lured to cities in search of a livelihood.

Working Together towards a Common Goal

"Traveling is learning," according to an African proverb. Sierra Leone and other at-risk countries stand a real chance of forestalling similar disasters by adapting the above solutions that have been successfully applied by their counterparts across Africa. EBAFOSA, through its strategy of innovative volunteerism, offers an opportunity for country stakeholders to convene their respective capacities for mutual partnerships towards a common end. The solutions are known, and we have the means to implement them. Let's rise and act in the best interests of Africa's present and future generations.

This is how the resilience of the African people, who have endured a fair share of suffering, will count.

Chapter 5

AFRICA'S OPPORTUNITIES UNDER THE CHANGING CLIMATE

From Silos to Creating Wealth through Systems Thinking

"Every adversity carries with it the seed of equal or greater benefit."

This is a truism I have come to appreciate in my own life. While we do not choose the circumstances of our birth, we do choose how we respond to them: either actively with boldness, courage, and hope that we can do something to improve our lives and take the right steps, or choose to do nothing. In which case, we have chosen to banish ourselves to mediocrity. I believe this is not how the creator, who endowed us all with the gift of intellect, the highest of any order of species, intended for us to live this life He gave us.

At this point, I am inspired to share another common saying across the savannahs of Africa, where the lion, the king of the jungle, reigns supreme. The saying is thus: "No matter the economy of the jungle, the lion can never eat grass; it is not pride, it is just who the lion is."

Analogising this to ourselves as human beings, the crème d la crème of all created life, way above the lion in creation's hierarchy, it is simply against human nature to live in mediocrity.

Across the globe, Africa is inadvertently portrayed as a continent of adversity. This only makes sense if your thought process is fixated on portraying problems but paying lip-service to the quest for solutions.

As a matter of fact, when we talk of malnutrition and infant mortality, these global statistics are significant: two hundred million Africans are undernourished, and 50 per cent of mothers are unable to feed their children to live beyond the age of five.

On hunger, over 240 million people in the continent go to bed hungry every day. Compounding food insecurity is ecological degradation. The depletion of ecosystems that are the foundation of food production through goods and services like water, healthy soils, and pollinators costs the continent $68 billion annually, coupled with losses of up to 6.6 million tonnes of potential grain harvest, capable feeding up to 31 million people. Furthermore, one-third of all food is lost as post-harvest losses amounting to $48 billion annually.

Turning to youth unemployment, with 60 per cent of youth being unemployed and ten to twelve million of them joining the labour market to compete for a fraction of the jobs (just three million formal jobs), Africa once again stands out. Related to this is illegal immigration, where Africa's most sovereign resource, its youth, risk life, limb, and dignity to cross the Mediterranean in search of illusionary greener pastures overseas. More than thirty-five hundred died in 2015 alone in the perilous Mediterranean. An unprecedented 70 per cent of illegal immigrants leave Africa for economic reasons.

While the need to create opportunities for millions is urgent, Africa's economies stand out for low productivity (an estimated 2000 per cent lower that of developed regions), primarily due to lack of value addition to commodities— that is, a lack of processing the raw commodities, for which the region holds a comparative advantage, into higher-value finished goods that fetch more in the market. As an example, in just one year, Africa's trade balance with China (its largest trading partner) moved into deficit of about $34 billion due to a fall in the price of many African commodities, such as oil, copper, iron ore, and cocoa. This loss is due to a focus on primary production. The continent earns a mere 10 per cent of the total extractable value from its agro-value chains due to low value addition. In the cocoa value chain, where the largest producer is in Africa, out of the over $100 billion in revenue made from chocolates alone, Africa receives a dismal 2 per cent, meaning it loses $98 billion of potential revenues and jobs.

Related to this low productivity is energy poverty, with 620 million (about 67 per cent of the population) without access to electricity. About ten million medium-sized enterprises lack access to electricity, and available electricity costs three times more than in the United States and Europe and is fraught with frequent shortages. Cumulatively, energy bottlenecks cost African economies up to 4 per cent of GDP annually, undermining sustainable growth, wealth creation, and investment towards unlocking urgently needed incomes and jobs.

On socio-economic inequalities, Africa is considered the second most inequitable region in the world. While the number of dollar millionaires has increased twofold since the year 2000, those living on less than US$1,25 a day has increased from 358 million in 1996 to 415 million in 2011. Over the past fourteen years, the number of high-net-worth individuals in Africa has grown by 145 per cent, much higher than the global average rate of 73 per cent. On the flipside, in comparison, while poverty has been reduced globally, it remains widespread in sub-Saharan Africa. Under the status quo, this scenario is projected to continue. While the number of African millionaires is projected to increase by 45 per cent by 2024, most of the world's poor will be living in Africa by 2030.

Overshadowing these challenges with a compounding effect is climate change, which threatens to shrink the economies of developing countries (most of which are in Africa) by a whopping 75 per cent. The African continent needs a minimum of $50 billion annually by 2050 to safeguard against this epic. Where will the money come from?

But elucidating the problem is not enough. More focus and attention should be paid to the quest for solutions. And this cannot be achieved through the usual run-of-the-mill approaches. Where sectorial actors constrict themselves to operating within their comfort zones, it will take alternative approaches, a systems approach that leverages on every sector, every skill in the continent, applied in complementary collaborations, and every challenge packaged as an opportunity towards a common end of maximising Africa's competitive advantages.

From Job Creation to Wealth Creation: A Change of Narrative

Africa's ballooning youthful population is projected to reach 830 million by 2050, with over 300 million joining the labour markets by 2035. Some have christened this scenario a ticking time bomb. But in the face of a rapidly expanding middle class, where consumer spending is projected to reach $1.4 trillion by 2020, business-to-business spending $3.5 trillion by 2025, and food markets $1 trillion by 2030, these youths display the potential for a demographic dividend for the continent. When harnessed in a complimentary manner, their skills, talents, interests, energies, and initiatives represent Africa's most fundamental capital to invest and unlock wealth opportunities in an increasingly affluent, consolidated market. And signs are all over on the potential of these youth. Just as an example, *Forbes* magazine's 2017 call for "most promising young entrepreneurs in Africa" attracted over 250 entries from twenty-three countries. Out of these, a total of thirty enterprises run by youth under thirty-five were nominated, with annual revenues of $100,000, a significant indicator of the high calibre of youth in the continent. Well-capable of remarkable enterprises that will transform Africa and ignite its economies. Add on to this the internet and mobile revolution sweeping Africa. Where tech-savvy youth in the continent have limitless access to information, knowledge, and networking capabilities that were unheard of just decades ago, prospects of an African demographic dividend are real. We can jump start this paradigm by leveraging on what we know and what is already ongoing in the continent.

I'm talking about energy poverty, where regardless of current challenges, most of the continent is blessed with 365 days of sunshine a year. A mere 0.3 per cent of the sunshine in the Sahara can supply nearly all of Europe's energy needs. I am talking about an isolated approach to electrification and energising the continent, where sixty energy initiatives in the continent, representing up to $30 billion in investments, are focused on the supply side, enhancing access to power, but this misses the point. Because considering that energy is an enabler of socio-economic development, indicators need to measure how energy investments have accelerated socio-economic transformation—not an arbitrary measure such as enhancing access. For instance, decentralising electricity by harnessing the abundant solar power to agro-industrialisation, we get a much higher return—an estimated $98 billion extra just from one crop value chain. This translates to not only social returns such as lighted houses, but economic and financial returns in form of enhanced incomes at the community and macro-economic scale, as well as mitigation of carbon where clean energy is prioritised. The implication therefore is to maximise returns; energy investments in the continent should be contextualised to envisioned end uses, where the priority should be maximising economic, financial, social, and environmental returns—not arbitrarily enhancing energy access and supply.

I am talking about climate change, which threatens productivity of the most critical sectors of Africa's economy. Key among them, agriculture employs a majority, an average of 64 per cent in Africa, yet is threatened by up to 40 per cent productivity losses, translating to not only lost food but livelihoods, as well. Targeted efforts to implement the Paris Agreement provide an opportunity to tackle this challenge. For example, leveraging ecosystem-based adaptation approaches will meet adaptation objectives under the Paris Agreement and increase yields by up to 128 per cent, boosting the income of farmers at a lower environmental and financial cost under the changing climate. It will reverse land degradation and put Africa on course to recoup over $60 billion lost annually due to land degradation, which is more than enough to cover the $50 billion Africa needs for climate action by 2050. Upscaling clean energy will mitigate against worsening climate impacts as called for in the Paris Agreement; it also represents an opportunity for sustainable agro-industrialisation to create wealth opportunities. However, this can only happen where the clean energy is tagged directly to powering agriculture products.

This exemplifies the continent's opportunity in adversity, aptly captured in an insightful African proverb: "The worlds of the elders do not lock all the doors; they leave the right door open." Thus, Africa's open door is two way.

First are its people, Africa's human capital, including the majority youthful population. The skills, talents, energy, passion, and networks of its people—young and old alike—adequately harnessed and optimally deployed, represent a priceless resource that money can't buy. They represent the continent's most sovereign capital.

Second is maximising productivity of its catalytic sectors. These sectors are economically inclusive, and the continent holds a comparative advantage in its resources here. Most importantly, these sectors can realise leading socio-economic priorities—especially food security, creation of income opportunities, and expansion of macro-economic growth for real wealth creation—alongside mitigating carbon and enhancing ecosystem resilience in line with continental climate objectives under the Paris Agreement. These climate objectives are commonly referred to as Nationally Determined Contributions (NDCs). Accordingly, farming strategies based on clean energy and ecosystem-based adaptation stand out as Africa's catalytic sectors, underscored by the AU Agenda 2063 and most recently by the African Ministerial Conference on the Environment (AMCEN) through the Ministerial Decision on Investing in Innovative Environmental Solutions.

On economic inclusion, agriculture is the most accessible economic sector, employing a majority of Africa's workforce (an average of 64 per cent across the continent and as high as 90 per cent in some countries). Maximising its productivity, therefore, means enhancing income and economic opportunities for the majority in the continent. Accordingly, the World Bank reports that in Africa, a 10 per cent increase in crop yields translates to approximately a 7 per cent reduction in poverty. Growth in agriculture is at least two to four times more effective in reducing poverty than in other sectors. EBA approaches are potentially cheaper, more accessible, and compatible to indigenous approaches used by Africa's smallholder farmers, who produce up to 80 per cent of the continent's food.

On comparative advantage in resources, 65 per cent of the world's uncultivated arable land and 10 per cent of its inland fresh water resources are in Africa. Beyond these natural resources is a three-hundred-million-strong middle class demanding more value added and differentiated agro-products; the food market is projected to be worth $150 billion by 2030. This represents a significant domestic consumer market for local agro-industries. Beyond this generic market, Africa can leverage on its competitive strengths in specific high-value and climate-resilient crops and tap into these value chains as strategic. Of note is cassava—a high-value, climate-resilient crop whose largest global producer is in Africa. This crop has unique market and health qualities that make it a high-value crop globally. Cassava is gluten-, grain-, and nut-free, as well as a vegan and paleo carbohydrate. This makes it a leading allergen-free food, whose global market is worth $23.5 billion. Cassava can be processed into over three hundred marketable products and by-products. The global gluten-free market alone is projected to be worth $7.59 billion by 2020. This is an indicator of the market advantage of allergen-free foods. In addition to tapping into the generic markets, Africa is positioned to also lead in the global allergen-free foods market.

On energy, Africa is endowed with diverse clean and renewable energy resources, where in addition to the best solar resource in the entire planet, as earlier noted, hydro's annual potential is estimated at 1852TWh, wind at 110GW, and geothermal at 15GW. By leveraging these resources strategically and deliberately to power-hungry industries—especially in the agro-sector, which offers the most competitive advantage—the region can establish global competitiveness and create much-needed wealth opportunities.

On meeting climate and socio-economic aims simultaneously, focused investments to maximise productivity of these sectors will meet leading socio-economic development priorities of actualising food security, enhancing income opportunities, and expanding macroeconomic growth, simultaneously with mitigating carbon and enhancing ecosystem productivity to meet the continent's climate objectives. To maximise this productivity, development in these sectors needs to be considered as complementary and not isolated, as classically approached. This amalgamation will potentially maximise productivity of agriculture by cutting PHLs largely driven by lack of power. By this reverse, the $48 billion lost as PHLs and recoup $35 billion currently spent importing food to translate to an injection of $83 billion into the regional economy. It will incentivise use of EBA to ensure up to 128 per cent yield increases under the changing climate, recoup food lost due to degraded ecosystems, and reverse loss of $68 billion annually driven by degraded ecosystems. Amalgamation will also maximise productivity of clean-energy development by diversifying application beyond domestic use in lighting households to include productive use in powering the agriculture industry. This will ensure a higher return on investment—going beyond traditional social returns to include direct financial returns in incentivising a $1 trillion agro-industrial economy by 2030 and unlocking up to seventeen million jobs.

Engaging Africa's human capital in maximising the productivity of these catalytic sectors represents its open door. It is how the continent will benefit from its seeming adversities.

And the formula to engaging this open door calls for a paradigm shift to one premised on collectivism rather than individualism, where the skills, experiences, talents, and initiatives of diverse complementary stakeholders—state and nonstate, individual, and institutional—are engaged in mutual partnerships towards a common goal: bridging policy and operational gaps towards establishing sustainable EBA-driven agriculture and clean energy-powered industrialisation. This is Africa's catalytic area and the engine of development, a break from traditional approaches premised on upfront financing of isolated development projects.

This is how Africa will divest from silo approaches to embrace integrated systems, where wealth and an accumulation of capital—both human and material—is realised as the basis for food security, creating income and jobs, expanding economies, and building climate resilience. The instruments to unlock this paradigm are already in place.

So let's get to work and turn Africa's adversity to opportunity.

Chapter 6

STATUTORY INSTRUMENTS TO UNLOCK THESE OPPORTUNITIES

Multilateral Agreements: Aligning to Africa's Advantage

One of my favourite African proverbs posits that "a large chair does not make a king."

Implicit in this saying is the fact that potential will remain unachieved unless steps are taken to harness it.

Countries across Africa are signatories to some of the most transformative and progressive, high-level global and regional development decisions and plans. By their very design, these decisions are intended to accelerate socio-economic transformation and achieve climate resilience. These twin issues are a top priority to guarantee the very existence of Africa well into posterity. But drawing their benefit is not preordained. It will take determined and strategic implementation, targeted at maximising productivity of the region's catalytic sectors and strengths, as earlier discussed.

Enabling Policies: The Biggest Driver of Change

From markets to technology transfer to financing, there already exist high-level policy levers for African countries to create an enabling policy environment for capital accumulation and wealth creation. This is the basis for accelerating Africa's socio-economic transformation and building climate resilience.

On markets, three multilateral agreements—the World Trade Organisation's (WTO's) Nairobi Package, the Yamoussoukro decision on creating a liberalised intra-Africa air transport market, and the African Union's (AU's) visa-free passport to allow for free movement of people across Africa—all provide a basis for consolidating the continent's products, services, and labour markets.

Internationally, during the Tenth Ministerial Conference of the World Trade Organisation in Nairobi, Kenya, member states moved a step further in levelling global trade. Captured in the historic Nairobi Package was a series of six ministerial decisions on agriculture, set to primarily benefit poorer member countries. Among the most significant is the commitment to abolish agriculture export subsidies, a significant step in levelling global agriculture markets to enhance Africa's opportunities to export competitively to more affluent foreign markets.

At the domestic level, increasing intra-Africa trade (currently the lowest globally at about 15 per cent, compared to 54 per cent in the North America Free Trade Area, 70 per cent within the European Union, and 60 per cent in Asia) is critical to harness the growing domestic market, projected to exceed $1 trillion by 2020.

Open Skies to Boost Trade

Efficient transport connectivity, especially air transport, is critical to harness this regional market and create additional livelihood opportunities. But this is being hindered by restrictive regulations and protectionism by some countries. The result is below par competitiveness in intra-Africa routes, leading to high costs where on average, flying within Africa can cost 200 per cent more than in Europe—a big market disincentive. This is among reasons that even with 12 per cent of the global population, Africa accounts for less than 1 per cent of the global air service market, with significant lost trade opportunities. For example, with open skies between just twelve key African countries, the resulting increased connectivity could create over aa hundred and fifty thousand jobs and add up to $1.3 billion to Africa's GDP. 100 per cent open skies across the continent would double or triple these benefits as the continent moves to maximise on its domestic market.

As a solution, the Yamoussoukro Decision, signed off by forty-four African nations, provides high-level policy levers towards deregulating the intra-Africa air transport. Follow-up implementation is on course with the launch in January 2018 of the Single African Air Transport Market initiative by the AU. This is a step forward from concept to implementation and covers key aspects of competition, consumer protection, and dispute settlement, among key safeguards to ensure efficient operation of the envisioned single market.

Seamless Cross-Border Movement: Buttressing a Consolidated Labour Market

In boosting intra-Africa trade, closely related to open skies is open borders.

African countries have the most visa requirements in the world. Only eleven of fifty-four countries, a mere 20 per cent, offer 100 per cent free access to all African citizens. This contrasts with the EU, where citizens enjoy 100 per cent freedom of movement in each other's territories.

This scenario restricts how labour can be deployed to address workforce deficiencies. Integrating Africa's workforce could potentially reduce brain drain, the emigration of talent out of Africa, while also allowing for its seamless deployment. Africa is reported to have lost a third of its human capital and continues to lose its skilled personnel at an increasing rate. With seamless movement of people across countries, the brain drain many African countries experience could become rather the transfer of talents across borders. This would ensure availability of an adequate, quality workforce, critical to accelerating the socio-economic transformation and building climate resilience, which we urgently need.

For example, an unemployed nurse from Ghana could earn a decent living in Liberia, enhancing family income while contributing to better health services in the host country to enhance social development. A young graduate from Tunisia's technical schools could find a decent job in the clean-energy industry in South Africa instead of venturing outside the continent. In both cases, human skills, talents, energy, passion, commitment (what constitutes Africa's most sovereign capital) are retained to create wealth and transform socio-economic resilience.

Once again, there are already high-level policies to turn seamless cross-border movement into reality. The AU's continental passport scheme will offer visa-free travel for African citizens within their own continent by 2020. Implementation should be quick and swift to benefit current and future generations.

The Paris Agreement: An Opportunity towards Maximising Catalytic Sectors

On 12 December 2015 at the Conference of the Parties (COP) 21 in Paris, 195 nations agreed to the historic Paris Agreement to combat climate change and transition to a low-carbon, sustainable future. Implementation of this agreement has been gathering increasing momentum, buoyed by the unwavering commitment to full implementation demonstrated by both state and nonstate actors at the highest levels. The cumulative implication is that the train of implementing the Paris Agreement long left the station.

With implementation gathering steam, Africa has forty-four states ratifying the agreement by February 2018, representing an over 80 per cent ratification rate (the highest of any region). This leadership is a demonstration that the region recognises the opportunities inherent in the agreement. One of the indicators of their priority to implement this agreement is their Nationally Determined Contributions (NDCs). A majority prioritise climate-proofing developments in economic sectors fundamental to the region's socio-economic transformation. Key amongst these is agriculture and energy, as well as restoration of ecosystems where EBA applications stand out. These stand as foundational in establishing the paradigm of sustainable agriculture, powered by clean energy, as the engine for accelerating socio-economic transformation and climate resilience in Africa.

Tapping into COP21 to Maximise Catalytic Area

Considering the Paris deal, such a paradigm will satisfy provisions of the agreement, including article 7 on adaptation, given that EBA is a climate adaptation technique. It will also align with articles 2 (on low emissions food production) and 4 on mitigation (especially through scaling clean energy), while also supporting the implementation of article 3 on NDCs. It is safe to conclude that operationalising Africa's catalytic area of establishing EBA-driven agriculture with clean energy to power agro-processing enterprises automatically drives implementation of the climate agreement. This provides the locus standi within the agreement to tap into opportunities in the deal as direct enablers of wealth creation for accelerated socio-economic transformation through sustainable agriculture. Articles 9, 10, and 11 mandate developed countries to support developing countries with appropriate means of implementation (finance, technology transfer, and capacity building) to enable them to meet their obligations under the deal. In this case, Africa should focus on appropriate clean-energy systems that can power various levels of agro-processing. For instance, the Paris deal secured commitments by a global alliance to mobilise up to $1 trillion to finance renewable energy development. Within the above frame, Africa's priority in any agreement pegged to this provision should be dedicated to clean, renewable energy appropriate for powering agriculture as a strategic thrust of implementing the agreement.

The Finance Question: Towards a More Market Sustainable Financing Model

Sustainable and affordable financing is another critical area over which Africa is well covered in high-level policy levers—awaiting only implementation.

Financing climate-resilient, wealth-creating, socially inclusive socio-economic transformation in Africa entails astronomical sums: more than $1.2 trillion annually. For instance, a minimum of $25 billion is needed to achieve universal access to modern energy, $18 billion for climate change adaptation, and $210 billion for basic infrastructure, food security, health, and security. Notwithstanding these astronomical amounts, Africa cannot rely on traditional public assistance, including official development assistance (ODA), whose totality has dropped to a mere 1 per cent of all capital inflows into the continent. At the GDP level, ODA now accounts for only 3 per cent of continental GDP. The implication is that relying on international public finance alone is a risky strategy; Africa needs to diversify sources. The continent needs to leapfrog from traditional financing approaches and embrace innovative, market-driven financing, the basis of which has been laid in high-level decisions on financing development, primarily the Addis Ababa Action Agenda on sustainable financing. This is buttressed by similar high-level studies and provisions, including the UNEP Inquiry Report on designing a sustainable financial system and the Second Africa Adaptation Gap Report. Implied herein is the need to go beyond dependence on traditional public finance towards a more market-driven and blended financing model that combines international public finance, domestic sources, and private sector financing to form a composite, more sustainable financing compact.

Domestic Sources? Show Me the Money

Mention of domestic resource mobilisation in Africa leaves people with a sour taste. The frowns emanate from a closely guarded belief that Africa is too poor to self-finance. This is defeatist and unacceptable, especially now with rapidly dwindling international support. It is time we all woke up and smelled the coffee. In 2015, the UN's flagship Second Africa Adaptation Gap Report proved, through judicious number crunching, that Africa can raise up to $3 billion annually of new finance from domestic sources to finance climate adaptation. In 2017, the UN Economic Commission for Africa (UNECA) concluded the continent already contributes 20 per cent of its total present adaptation needs, estimated at $15 billion annually. This translates to another $3 billion annually, making a total of $6 billion annually from domestic sources to build climate resilience.

Beyond the above, reversing current losses represents another critical source of domestic finance for the continent. Recommendations on stymieing illicit financial flows, as articulated by an AU/UNECA panel, can see Africa recoup over $50 billion annually—with some estimates as high as $69 billion—to finance its development agenda.

In addition, the economic cost associated with ecosystems and land degradation in sub-Saharan Africa is estimated at $68 billion annually. Cumulatively, Africa loses an estimated $100 billion annually that can be harnessed as domestic resources to finance development.

Beyond reversing losses, an African central bank can spearhead establishment of market-driven sustainable financing for Africa's development. Three key functions of a central bank position it to drive such market-driven financing: credit control, management of cash reserves, and lender of last resort. The Sirte Declaration of 1999 establishes the high-level policy lever to push for creation of such a bank, which should be specifically dedicated for credit control and risk management of commercially viable loans to bridge financing gaps in maximising productivity of the continent's catalytic sectors. Shareholding of this continental central bank could be distributed amongst central banks in countries across the continent, which should then lead in raising a continental cash reserve for the bank.

With the above, it can be considered that Africa has what it takes to implement a more sustainable financing model, one that weans the continent off dependency on aid benevolence and development finance as a social undertaking to instil a model premised on sustainable market-driven finance. Where returns are not only social, but environmental, economic, and financial. This is the future of development finance, and Africa has what it takes to boldly take it on.

Dotting the Is and Crossing the Ts in Enabling Policy Levers

In tying the above enabling policy levers into a coherent policy narrative towards maximising productivity of Africa's catalytic sectors for wealth creation and climate resilience, 2017 stood out as a watershed year. There was an adoption at the continental and global levels of a potentially transformative policy lever that will position environment and climate action front-centre in Africa's push for accelerated socio-economic transformation. Building ecosystem resilience and mitigating carbon will ensure food security, create wealth opportunities to unlock jobs, and expand macro-economic growth, all through maximising the productivity of catalytic sectors. This portends good tidings for people and the planet.

At the sixteenth AMCEN, ministers of the environment unanimously adopted a ground-breaking decision on "investing in innovative environmental solutions to accelerate implementation of the Sustainable Development Goals and Agenda 2063 in Africa." This decision condensed into one coherent narrative, the key enablers to maximising productivity of Africa's catalytic sectors as discussed herein. One of the key areas covered was finance. Here, this decision talks of innovative financing, which establishes the trajectory to unshackle the continent from aid benevolence for progress towards leveraging on market-driven financing, where development finance is no longer premised as a social undertaking but as a financial investment capable of social, environmental, economic, and financial returns. This decision looks to policy harmonisation to create an enabling policy environment. It is driven by the realisation that tapping into the continent's catalytic sectors will take multiple ministries working in unison. This provision prioritises the bridging of ministerial silos for policy coherence across multiple relevant ministries, creating an enabling policy environment across government to maximise productivity of catalytic sectors as the engine of accelerated socio-economic transformation and climate resilience.

The decision also goes further to prioritise targeted education and capacity-building aimed at integrating the catalytic sectors and integrating them in a formal curriculum to ensure adequately capacitated human capital towards maximising productivity of the region's catalytic sectors. This strategy ensures the continent's growing and youthful population (wrongfully characterised as a ticking time bomb) becomes a demographic dividend.

These high-level policy levers, through continental leadership, were further scaled to the global level with the unanimous adoption at the Third United Nations Environment Assembly of a resolution on "investing in innovative environmental solutions for accelerating implementation of the Sustainable Development Goals." This meant that Africa's policy charge to leverage on its catalytic sectors as accelerators of socio-economic transformation and climate resilience had received global support. It doesn't get any better than this.

It was extremely refreshing to have led both processes from start. Given another chance, I would do it again. But I'm not resting. My team is not resting. This is only the first step in a thousand-mile journey to maximise productivity of Africa's catalytic sectors for accelerated socio-economic transformation and climate resilience. We are now moving to consolidate these gains and turn the promise of these high-level policy levers to the reality of food-secure homes, of gainfully employed youth, of expanding economies, of a climate-resilient people, and of a flourishing planet. Achieving this will take collectivism, not individualism. My team and I cannot do it alone. We need all of you to join us and work with your minds and hearts, using your skills, talents, and initiatives to bridge gaps towards this flowery end we deserve as a people. This is the essence of innovative volunteerism—a call to a change of attitude—where we look to our skills, talents, and initiatives: the essence of our very being, the solution to the challenges, and the sovereign capital to harness the opportunities that confront us. It's a call for duty to each one of us.

Comrades, let's get to work and make the large chair that is these progressive policy provisions turn Africa to king.

Chapter 7

INNOVATIVE VOLUNTEERISM: UNLEASHING AFRICA'S SOVEREIGN WEALTH

Power to the People: Engaging Africa's Sovereign Capital— Its People

An insightful African proverb reflects innovative volunteerism: "Cross the river in a crowd, and the crocodile won't eat you."

This seemingly amusing proverb appeals to a virtue that is universally shared across the continent: solidarity and unity of purpose. This is the gist of innovative volunteerism that you will encounter as you read along.

The late President Nelson Mandela could probably develop a full thesis from this singular proverb drawing from his experiences as the embodiment of the anti-apartheid struggle and his stewardship of the young rainbow nation of South Africa, soon after her independence.

He left us with countless other sayings. Among my favourites is that "we must use time wisely and forever realise that the time is always ripe to do right."

Never in Africa have these words rung as true as they do presently. The urgency to solve Africa's developmental challenges has reached fever pitch.

Intrinsically and simply, wealth is the accumulation of capital—human and material—which is the foundation upon which accelerated socio-economic transformation and climate resilience is built. Human skills, creativity, talent, and energy constitute the most sovereign, critical, and significant component of wealth globally. With human capital, one can generate material. A big damper to Africa's progress has been the lopsided prioritisation of material over human capital. This is the reason the Africa rising narrative was a nonstarter. The Africa rising boat, fuelled by rapidly appreciating global commodity prices, never sailed far. Once it ran into the turbulence and headwinds of commodity volatility, with prices of leading commodities like oil plummeting over 40 per cent in a single month, Africa's journey to economic freedom was doomed. This is a lesson for the continent not to prize material over its people, who are its most sovereign capital. A prudent approach is to invest such material resources into buttressing human capital to maximise inclusive sectors, in which sustainable agriculture is at the core. Not entirely consuming these resources. But this is a topic for another day, which has nevertheless been much discussed globally under topics such as the natural resource curse.

Back to our topic of interest here; as earlier elucidated, Africa is currently the youngest continent in the globe. This means it is potentially most advantaged to harness its human capital. And considering the prevailing challenges and opportunities, Africa's youth are at the brink of greatness for themselves and the continent. They are the embodiment of Africa's demographic dividend, the tip of the spear, the game changer to accelerate Africa's match to victory against lack, socio-economic regress, and climate vulnerability among its leading enemies.

The bigger question is how.

An illustrious son of Africa's transformation journey, having perceived the continent's very palpable possibility of steady and accelerated socio-economic transformation, boldly declared, "Africa will write its own history, and both north and south of the Sahara, it will be a history full of glory and dignity." Although Patrice Lumumba was laid to rest in 1961, under the most unfortunate circumstances imaginable, these profound words do not rest in his grave. But are alive and well in the hearts and minds of all Africans worth their salt. These words have haunted the continent for over five decades. Now more than ever, as socio-economic pressures and vulnerabilities sour, the urgency of their fulfilment beckons us all. The responsibility for working towards their fulfilment surely rests on the shoulders of each one of us—current and future generations. There can be no rest, until these prophetic words are fulfilled.

On the how question, the prompt answer is innovative volunteerism, which is premised on a very simple idea: engaging the entirety of Africa's human capital in its diversity towards a shared goal of maximising the productivity of the continent's catalytic sectors, as discussed in chapter 5. In this pursuit, this ideal calls for a paradigm shift well imbedded in African culture, that of divesting from individualism to embrace collectivism, banishing selfishness to take on selflessness, rejecting hopelessness for hopefulness. This is the philosophy and paradigm behind innovative volunteerism, where the skills, experiences, talents, networks, and initiatives of diverse complementary stakeholders (state and nonstate, individual, and institutional) are engaged in mutual partnerships that meet the respective business and organisational interests of these actors but converge towards a common goal: bridging policy and operational gaps to establish a sustainable industrialisation of the agriculture sector, powered by clean energy. Africa's strategic acceleration of socio-economic transformation and building climate resilience will ensure the continent's competitive advantage of a youthful population and superior resources is converted into a global competitive advantage.

To converge and convene the above towards coherently operationalising innovative volunteerism requires an inclusive framework. Africa has already given itself such a framework. It all started with a continentally commissioned task force study that I had the privilege and honour of coordinating. Its express mandate was to establish an optimal strategy through which Africa could leverage its often small-scale climate interventions as a seed to build large-scale, impactful actions that enhance food security. This study drew perspectives from many practicing experts across the continent—scientists, policy experts, government bureaucrats, academics, and private sector actors from diverse sectors. The study concluded by recommending an inclusive framework that would harness complementarities across the diverse sectors in Africa towards this goal.

This recommendation was quickly followed by actual implementation. In July 2015, less than six months from publication of the task force's findings, the African Union, United Nations Environment Programme, the African Development Bank, and the African Centre for Technology Studies facilitated the second Africa Ecosystem-Based Adaptation for Food Security Conference to deliberate on the study's findings and chart a pathway to its implementation. This conference drew more than twelve hundred experts from diverse disciplines—state and nonstate—as well as ordinary citizens; here, the Nairobi Action Agenda on Ecosystem-Based Adaptation for food security was unanimously adopted to establish the Ecosystem-Based Adaptation for Food Security Assembly (EBAFOSA), a framework for policy implementation. This was a giant step in fulfilling the task force's recommendations to maximise the impact of climate interventions in Africa.

It was truly fulfilling for me to rally our continent to speak with one voice in charting solutions to problems we share, regardless of borders.

But what is EBAFOSA, you may ask?

EBAFOSA takes centre stage in mobilising Africa's human capital in operationalising innovative volunteerism. It is an inclusive policy implementation framework that convenes complementary actors drawn from diverse sectors and academic disciplinary backgrounds at policy and operational levels—especially environment, agriculture, energy, ICT, industry, and administration. State and nonstate organisations, individuals, and institutions can build on their skills, expertise, and ongoing initiatives to develop mutually beneficial partnerships that meet their respective business and organisational aims but align to actualise the shared EBAFOSA objectives. These actors bridge policy and operational gaps to maximise the productivity of Africa's food systems by establishing agriculture processes that are ecosystem-based and powered by clean energy, with efficient ICT linkages to markets and supply chains as the strategic engine for wealth creation. Wealth creation is the imperative for the continent's accelerated socio-economic transformation and climate resilience.

The EBAFOSA partnerships to bridge implementation gaps draw from section 5 of the Paris Agreement and goal 17 of the sustainable development goals. Both underscore multiple-stakeholder partnerships as a practical means to bridge implementation gaps. EBAFOSA provides a common framework to convene these partnerships and achieve synergy amongst diverse actors. An added advantage of this mechanism is that it builds on a foundation of established ongoing work (rather that starting new initiatives from scratch). The result is diversified, with a lower implementation risk. Another advantage is a minimisation of duplicated effort. Cumulatively, this approach maximises the impact and efficiency of programmes.

EBAFOSA Modus Operandi

The EBAFOSA modus operandi entails collectivism through complementary actions, where multiple stakeholders—state and nonstate; individual and institutional—develop mutually beneficial partnerships using what they do, the skills they have, the talents they possess, and the networks they have built as the premium. These partnerships are designed to meet their respective business and organisational aims but align to actualise the shared EBAFOSA goal, that of bridging policy and operational gaps to maximise the productivity of the catalytic sectors. This is a prerequisite to solving the socio-economic challenges of food insecurity, unemployment, and economic stagnation while building climate resilience. These win-win partnerships among complementary actors to bridge the gaps mentioned above represent innovative volunteerism.

But for this to happen, we must connect the dots. Connecting the dots can only happen within an inclusive framework that convenes all these diverse complementary actors to develop their partnerships. This need is what resulted in the establishment of the Ecosystem-Based Adaptation for Food Security Assembly. Established collectively through a continental process, EBAFOSA is a country-driven, inclusive policy implementation framework that convenes multiple sectors of the economy to work in synergy to accelerate socio-economic transformation and climate resilience. EBAFOSA is therefore the means by which innovative volunteerism takes place across the continent.

As you can see, EBAFOSA is inclusive; everyone's skills and talent matter. It's an assembly of people motivated to act, not by the challenges but the opportunities Africa presents, a coalition of the willing founded on complementarity, not silos. This is where everyone's strengths are complemented by another's towards a shared end: to leverage and expand opportunities in working with nature and create wealth for themselves and for those yet to be born. It's where vibrant and youthful skills, talents, professional knowledge, creativity, and energy combine with professional experience to unlock wealth opportunities.

EBAFOSA is erected on selflessness and anchored on collectivism, where each and every one of us has a part to play—individually as well as through the institutions we represent—to realise a desirable shared end. With innovative volunteerism, what counts most is not who you know but your passion and commitment to drive transformational change through sustainable agriculture, powered by clean energy.

EBAFOSA Pillars

EBAFOSA partnerships are targeted at five EBAFOSA pillars. The foundational pillar is the linkage of clean energy to power processing of nature-based, ecosystem-based adaptation (EBA) agriculture products, as the basis of sustainable agro-industrialisation. The synergy achieved by integrating developments in these sectors—as opposed to conducting them in silos—is to maximise food security and income opportunities, simultaneously with mitigating carbon and enhancing ecosystem resilience to meet climate objectives. This approach incentivises countries to transition to low-emissions development.

This foundational pillar is then complemented by market enablers that include the following:

EBAFOSA Compliance Standards. These standards are applied universally across all EBAFOSA countries for quality and environmental compliance along the EBAFOSA continuum. This standard serves to consolidate a continental market of both fresh and processed produce that meet the threshold of EBA-Driven Agriculture & Clean Energy. Certified products can be marketed across the forty EBAFOSA countries in Africa. This certification therefore serves to harness the agro-market worth $150 billion across Africa.

Innovative Financing. The strategic underpinning of this pillar is the need for Africa's self-financing, pegged on the fact that donor fatigue is now a reality. The implication is Africa cannot rely on donor benevolence to finance development and climate resilience. But this gap opens an opportunity for alternative market-driven models to bridge it. Innovative financing is an alternative model based on market-driven financing of enterprises along the EBAFOSA continuum. It leverages on the principle of cooperatives/co-guaranteeing schemes and the risk diversification they provide to unlock affordable, market-driven financing. For this, multiple actors along the EBAFOSA continuum convene in a cooperative to mobilise their collective resources, diversify risks, and pool finances towards

> ➢ directly purchasing capital assets (e.g., various types of clean-energy systems), inputs, and services along the EBAFOSA continuum (where capital assets are involved, these can be used communally),
> ➢ using the pooled resources as security to enable them to leverage more finances from micro-finance institutions to acquire capital assets, inputs, and services along the EBAFOSA continuum as above, and
> ➢ specific to capital assets and other capital-intensive inputs, the co-guaranteeing schemes and cooperatives can work with suppliers of these inputs (clean-energy system suppliers, input suppliers, service providers, etc.) and micro-financers to develop products targeting financing of such capital-intensive assets.

Use of Information Communication Technology. The key issue here is developing mobile-eBAFOSA (M-eBAFOSA) in each country, which is a virtual EBAFOSA accessible on both mobile and desktop platforms. It operationalises all the five EBAFOSA pillars through a series of menus. At the operational level, M-eBAFOSA is about connecting the dots in each country on operational processes, products, and services along the entire EBAFOSA continuum. For example,

> Linking on-farm actors to extension and advisory service providers on both farm EBA and clean energy.
> Linking farmers to clean-energy actors (suppliers of technologies and advisory services) for various levels of clean energy-powered agro-processing solutions.
> Linking all these actors to finance service providers. Here, all cooperatives will be mapped as providers of finance services.
> Linking producers to EBAFOSA-compliant services. M-eBAFOSA provides an e-commerce platform where EBAFOSA-compliant produce is marketed and sold online across the country and the forty EBAFOSA countries across Africa. Cooperative members whose produce is certified will have their produce access a wider market.
> At the policy level, M-eBAFOSA offers a lot of data for policymakers. This provides officials with information that enables them to make optimal decisions on location of the agro-industrial zones in operationalising all pillars.

The above operational pillars are augmented by the following policy pillar.

Policy Harmonisation Task Forces. These are policy organs at the ministerial level to synchronise programmes from multiple relevant ministries for coherent and complementary strategies to enable all the above pillars to function optimally.

All the EBAFOSA pillars are driven through innovative volunteerism, which leverages on people's skills, networks, and ongoing actions in market-driven collaborations to bridge implementation gaps, as explained earlier. By doing this, EBAFOSA builds on ongoing initiatives and aggregates them for bigger impact, rather than launching new initiatives from scratch. It reduces the problem of silos in Africa, where different stakeholders end up duplicating efforts rather than working together for synergy, which has a bigger impact.

So what has EBAFOSA achieved through innovative volunteerism?

After just a year in action, innovative volunteerism has proven that it's not blind optimism. Multiple actors across the continent have taken to it like a fish to water. Whether we look at the coastlines of Nigeria, Benin, Côte d'Ivoire, the Gambia, Togo, Ghana, Liberia, and Sierra Leone in West Africa or decide to consider the savannahs of Cameroon and DRC in Central Africa and move down to the highlands of Kenya, Tanzania, and the Matoke hills of Uganda, all in East Africa. A similar story confronts us in the savannahs of Malawi, Zambia, Mozambique, and Zimbabwe in Southern Africa. The message is simple. Through harnessing the spirit of innovative volunteerism, EBAFOSA is generating waves of progress and driving socio-economic transformational change across Africa.

Examples are proving how EBAFOSA's critical pillars are leveraging on Africa's sovereign capital—its people. People are exercising their skills, and networks are tapping into the sustainable agro-industrialisation economy to create livelihoods for self, country, and continent.

At the policy level, EBAFOSA's Sierra Leone Policy Task Force is building on four ongoing policy initiatives across complementary ministries to establish tax concessions for agro-based industries in rural areas. Establishment of this tax concession policy is aimed at incentivising investment in clean energy-powered agriculture enterprises near farming areas. This will enhance community incomes and food security and also address climate objectives of mitigating carbon to meet Sierra Leone's NDC.

In the Gambia, the task force is devising a solution to one of the most critical challenges faced by the country (but shared continentally): illegal migration. Youth in the Gambia are migrating illegally to Europe in search of better livelihoods. To curtail this challenge, the Gambia Department of Immigration is working with complementary ministries— energy, agriculture, and industry—to develop a counselling curriculum for youth; this interagency task force taps into income opportunities in EBA-based, clean energy-powered agro-industrialisation.

Operationally, in Cameroon, EBAFOSA is catalysing partnerships at policy and ground levels to link off-grid small-hydro systems to power cassava and Irish potato processing; there are a variety of product lines linked to markets and supply chains using ICT mobile apps. This is not only offsetting carbon in energy generation and building ecosystem resilience by incentivising use of EBA approaches to meet Cameroon NDCs, but also creating income opportunities along the entire agro-value chain and ancillary chains of clean energy and ICT. A total of ten youth groups working in ICT, clean energy, and marketing have been engaged creating green jobs for approximately a hundred young people. More than five hundred women now have access to value addition services and as a result have cut their post-harvest losses (PHLs) to enhance their income stability and increase the community's food security.

There is a formula to engender an inclusive Africa and ensure collective progress and prosperity. Such pockets of success are a clarion call, encouraging us to build on them and create full-scale solutions. And it is our voluntary but targeted complementary actions that will establish these enablers. Let's not despise ourselves. As the African proverb goes, "If you think you are too small to make a difference, you haven't spent the night with a mosquito." If we don't despise ourselves, our collective actions will move mountains. EBAFOSA provides you with that opportunity to make a difference now and into the future.

A difference like that being made by Muhammed Taal in the Gambia. After fourteen years in Europe, doing different jobs to survive, he returned home with little to show for it. But thanks to his skills in manufacturing compost manure and creating a livelihood in it, Mr Taal is now gainfully engaged in the Gambia. Through EBAFOSA, he has partnered with livestock farmers to harvest manure and groundnut farmers to harvest shells, using these raw materials to produce and sell bio-fertiliser. His enterprise is now mapped as a supplier of organic fertiliser on M-eBAFOSA so it can access markets across Gambia. Through the EBAFOSA compliance standard, which recommends bio-fertiliser as part of standards compliance, demand for the fertiliser is enhanced, which enhances his earnings. This enterprise not only improved Mr Taal's livelihood, it contributes to meeting the Gambia's climate objectives as expressed in its NDCs. He is also training youth in this enterprise to replicate this success.

It's a story of young students of St Kizito High School in Uganda, who are developing practical skills in briquette production, smart agriculture, and waste management; they are also supplementing their technical skills with business and entrepreneurial training. This initiative successfully transformed St Kizito High School from using firewood to using their fuel briquettes, saving up to seven trucks of firewood per term and twenty-one trucks per year. These entrepreneurially promising youth are Africa's future; it's quite encouraging.

So the writing is on the wall. It's no longer tenable to keep talking about Africa's great potential; rather, it's time to roll up our sleeves, engage in innovative volunteerism, and fulfil the continent's potential for the benefit of present and future generations. Just as colonial oppression drove our founding fathers to launch the struggle for independence, today the urgent need to accelerate the continent's socio-economic transformation beckons all citizens of the continent to action. The time to do the right thing and act is now. EBAFOSA provides the means to engage; innovative volunteerism is the method.

Chapter 8

CALL TO YOUTH

Realising Africa's Demographic Dividend: Where the Mind Goes, the Man Follows; A Call to the Youth

"It is the young trees that make up the forest." With up to 60 per cent of the population being youthful, Africa is the richest continent, more so considering that human capital is the most important component of global wealth.

The big question is, how can this great potential, the over two-hundred-million-strong youth, be harnessed? The answer is in mindset change among the young (and young at heart) on this continent, those ready to do things differently—within the confines of ethics beyond reproach—for the good of present and future generations. This is the urgency of our time. Because once we have the minds of our youth thinking right, they will act right. They will act passionately to build wealth for the continent through using their skills, talents, energies, and initiatives to achieve socio-economic transformation. They will be guided by the broader picture—the collective benefit of themselves, of present and future generations, and of the planet—not myopic self-interest.

A bulk of the responsibility for mindset change lies within the individual. The values, personal initiative, passion, drive, and determination we hold as individuals is paramount because to succeed, innovative volunteerism calls for a youth ready to work with their heart and mind.

Young People Have Tools to Drive Innovative Volunteerism

Way back in 2013, the UN reported that there were 650 million mobile users in Africa—probably much more now. Thanks to the internet, today's tech-savvy youth have greater access to opportunities than the generations before them. The internet, mobile technology, and social media have revolutionised how we network and make our voices heard. The tech-savviness of young people represents an opportunity to enrich policy engagements.

Along with regular citizens, political leaders have rapidly joined social media platforms, using them to engage the masses. In turn, this is an opportunity for the youth to engage political leaders objectively regarding policy positions. Internet-enabled phones and social media provide the youth with a golden opportunity to actively engage with political and policy leaders by lobbying for policies that drive the catalytic areas as a basis for wealth creation. Promote good governance as the basis for sound wealth management. This is an aspect of innovative volunteerism that organised youth can quickly tap into.

Beyond activism, every skill is invited in the exercise of innovative volunteerism. How? By building targeted partnerships with complementary actors that meet our respective business objectives but align to actualise a shared, grand continental mandate of wealth creation through the catalytic sectors. And EBAFOSA provides us all the platform and framework to forge such mutual partnerships.

Why? Because Africa's success is our success. As you contribute your legal knowledge to set up an agro-industry, you get yourself a client while creating millions of jobs along the agro-value chain and in ancillary supply chains of clean energy and ICT. You contribute towards combating youth unemployment in your country and Africa, even as you expand your clientele base.

As you work your farm, EBAFOSA partners you with advisory service providers to implement nature-based EBA approaches. You get access to standards to improve your products so they fetch a bigger market—spanning as far as forty countries. And in the process, you not only increase your earnings, but contribute to ending hunger beyond your borders and to combating climate change, using more of EBA.

As you study to become an ICT or clean energy professional, through EBAFOSA you get the opportunity to work with established professionals, perfect your technical skills, and become a competitive, employable professional while contributing to establishing sustainable industries that will expand our economies.

As you run your clean-energy enterprises, EBAFOSA gives you an opportunity to partner with finance actors for affordable finance to expand your agro-industrial applications. And it goes further as you get mapped into EBAFOSA as a supplier of these technologies to expand your market. In the process, you not only enhance your business earnings, but contribute towards making Africa a global leader in clean industrial applications, creating clean jobs.

As a banker, EBAFOSA offers you an opportunity to be part of risk-sharing facilities that not only expand your loans income portfolio, they lower the default risk to enhance your bank's profitability. And in the process, you contribute towards job creation and economic expansion, which both serve to guarantee you longer term clients.

Policymakers are not left out. As a policymaker, EBAFOSA provides you with a platform to work with colleagues from complementary ministries to maximise the impact of your policies for grander impact—that of industrialising your country and Africa. This is EBAFOSA innovative volunteerism at work, where everyone's skills and talents find a voice and find relevance in driving development.

Granted, it will not be easy—especially for youth who are just starting out, perhaps straight from college or high school. But your skill, determination, and effort are all it takes. And the reward is worthwhile. Let me encourage you with the story of Pascoal Silverio, a Mozambican youth, just like hundreds of millions across Africa. Pascoal was born in 1990. I was introduced to him by a colleague, who met him in the small village of Pemba in 2013. Back then, barely in his twenties, he was translating for a meeting. Young as he was, he amazed me with his listening skills and his ability to simultaneously translate words from Portuguese to English and vice versa. But his background was far from glorious. Partially blinded as a child, Pascoal encountered difficulty studying and interacting with his sighted schoolmates. As if that were not bad enough, he lost his mother, his source of emotional and physical strength, at a young age. His life seemed doomed. It seemed he had hit a brick wall with nowhere to turn.

But by God's grace, and driven by his dream to be somebody in life, the partial blindness forced him to sharpen his listening skills and excel in school to qualify for university. He didn't stop there. After graduating, he volunteered his knowledge, skills, and talents in the relevant sectors aligned with his line of training. This allowed him to network, meet industry leaders, receive mentorship, and refine his technical skills. He showcased his talents, skills, and knowledge, positioning himself for a quick rise up the ranks.

Now in his midtwenties, Pascoal has had four different jobs, leveraging experiences and networking in one job to get a subsequent better job. I would not be surprised to see him on TV, sharing the story of his rise to become an iconic African entrepreneur. Whatever he chooses, he is on track.

If Pascoal is making it despite his physical challenges, how much more can you, the abled bodied youth of Africa?

This is the power of volunteerism towards a bigger goal, the power of focus, which is what I am calling you to embrace and act on. Package your competences and dreams, whatever they may be, be they in science, in history, in engineering, in the performing arts, in politics, in entrepreneurship of any field, towards the shared objective of complementing the actualisation of sustainable agro-industrialisation in Africa. By so doing, in the next ten years, you will be part of the $1 trillion agro-industrial sector projected for Africa: not only creating it, but savouring and enjoying it. This is the way to go. And the EBAFOSA framework provides a ready platform for engagement, to ensure your actions will not be in a vacuum.

You may ask, "What will I gain right away?"

Like Pascoal, through volunteering the skills, knowledge, and resources at your disposal towards this bigger goal of industrialising Africa's agriculture, you get to network and interact with industry leaders. You showcase your skills, get mentored to refine them, and earn practical hands-on experience that positions you competitively in the labour markets. Let the never-give-up attitude of Pascoal energise all of us so that through innovative volunteerism, we can simultaneously develop ourselves and contribute substantively towards wealth creation in our countries and continent.

An Education Tailored to Drive Innovative Volunteerism

Beyond the youth themselves, who bear the greatest responsibility, policymakers also have a critical role to give our youth the best possible start. Africa's education policies need to be reformed and centred on solving the continent's challenges—through the strategic thrust of the catalytic sectors as discussed in chapter 5, while technically benchmarking against global best practices. Education should add on its rigorous theoretical basis, a targeted alignment to solve the continent's development challenges. Curricula should equip students to do so.

There is also a need to invest in compulsory quality universal education through the following measures:

Staff Rationalisation Policies: There is a need to ensure all institutions of learning are equally staffed with adequately trained personnel. Policies that practice periodic countrywide staff audits will go a long way in correcting staffing imbalances and ensuring quality staff to all students. Lessons from South Africa indicate that a critical success factor of rationalisation programmes is the budgetary allocation for hiring new staff whenever deficits are revealed by the exercise.

Education Finance Reform: National governments and the private sector should come up with innovative scholarship and grant policies that ensure financing is targeting programmes aligned to maximising catalytic areas. Attracting students to these programmes and monitoring their development will ensure adequate and skilled personnel to confront country-specific challenges. Private-sector participation through the offering of corporate scholarships fosters the sustainability of such initiatives.

Policies Restructuring Tertiary Institutions' Research Capacity: There is also a need for policies that promote continuous improvement in the quality of training and research. Institutions of higher learning should institute policies that increase the capacity of the academic staff. Examples include policies offering periodic grants for fellowships to top universities across the globe; policies to enhance institutional collaborations with benchmarked global universities; and policies enabling partnerships with the government, private sector, NGOs, and other research-oriented academic institutions to address specific challenges.

From the youth themselves to policymakers, we have the how, so let's make it happen.

Chapter 9

AFRICA: A SHINING CITY ON THE HILL

The "Africa We Want" Enshrined in the AU's Agenda 2063—Yes, It's Possible

Part of the mantra of the AU Agenda 2063 reads "optimize the use of Africa's resources for the benefit of all Africans." This sentence is strikingly appealing. It speaks of the African paradox that confronts the entire globe; how is it that the continent most endowed in natural resources is the poorest, its people most vulnerable? How come all the intentions of good will towards the continent—in the form of some of the best development strategies and plans ever formulated—end up to naught? We have these great aspirations of the AU Agenda 2063, which are a strategy unto themselves, yet there is talk of an African Marshall Plan. Is another strategy what Africa needs?

Well, we can say yes and no. Yes, if the strategy is about implementing what we already know—which is a lot. A big no if we are getting more paper strategies. Africa needs to kick to the kerb the vicious cycle of endless strategies that are so good on paper yet never get implemented. Africa needs to move from paper to implementation, to consolidate its strengths around the catalytic areas and implement actions with speed.

We can learn from several Asian tigers. South Korea is a typical case. This mineral-barren country roughly five times smaller than Kenya has an economy estimated at fifteen times the size of Africa's combined economy—the totality of fifty-four countries in the continent. Its per capita income—the wealth produced annually divided by its population of over fifty million—grew 275 times from $82 in 1960 to $26,204 in 2013.

But it was not so, some years back. In 1962, just a year before Kenya, where I am writing from, broke loose from the grip of colonial rule, and two years after some seventeen African countries—the largest bulk in a go—had broken loose from the grip of colonial rule in 1960, South Korea, then a rickety Asian nation, faced with dry coffers, was laying the cornerstones of what would become the economic powerhouse we know today. Back then, Korea's backwater economy was neck deep in sickly agriculture that accounted for 75 per cent of its output. Agriculture was its most critical sector then, just as is the case for Africa today. Productivity was at an all-time low. So bad was the situation that the Korean population faced acute starvation, prompting goodwill from nations as a matter of urgency. In 1963, Kenya gave out a $10,000 loan and provided relief food to South Korea. The loan was later repaid.

But more exciting is the lesson we gain from South Korea's rise. Its meteoric rise was based on maximising strategic industries of textile and clothing for exports as the main economic thrust. This was achieved by converging key enabling sectors towards complementing productivity maximisation in this strategic area. This was how South Korea's rise began.

The country also marshalled resources around training, alongside research and development for an innovative, creative economy. The state invested heavily in engineering, sciences, ICT, and technical courses for human capital accumulation. The country prioritised incentives to boost competitiveness of their exports, drawing from their strategic area of textiles and clothing. For example, in the quest to ramp up exports and build up forex reserves, the government would entice industries into closing sales outside the country in return for corporate income tax waivers and tariff exemptions—a very enticing incentive package. Africa can target similar incentives to fuel the growth of its catalytic area of clean energy-powered agro-industrialisation.

Basically, South Korea—without the advantage of Africa's natural resources base—used the textile and apparel industry to transform the fortunes of the country. The export-led strategy leveraging on textiles and apparel and its human capital grew Seoul's exports thirty-six times from $22 million before the development plan to $835.2 million in 1970—less than ten years from the crisis year of 1962.

The lesson is simple: A rolling stone gathers no moss. Africa must focus on industrialising the areas of its comparative advantage—in this case, the catalytic sectors as elucidated in chapter 5. Lessons from South Korea and the entire developed world show that industrialisation still remains the shortest route to national wealth creation—the prerequisite for accelerated socio-economic transformation and building climate resilience.

Drawing parallels with South Korea, which without benefit of a natural resource base or a domestic market grew its exports thirty-six times in less than ten years, Africa's performance should be much faster. With a potential $1 trillion domestic market to start from and sharpen its competitiveness, a relatively well-structured service sector, be it ICT or finance, the benefit of rapid technological deployment the world over, and a natural resource base, Africa's rise to a leading global actor in sustainable agriculture powered by clean energy should be much faster. With this determined focus, it should not be long from now that African countries develop their own product brands drawing from the catalytic area that are sold competitively across the continent and globe.

This is the essence of innovative volunteerism, where human, material, and natural resources as well as intangibles like policies line up in unison towards a shared goal, that of maximising productivity of a strategic area, with the view of catapulting the country to global competitiveness. This is how wealth, a prerequisite to accelerating socio-economic transformation and climate resilience, is built. It will be no different for Africa.

CONCLUSION

Making Africa Truly Great

God is no respecter of persons; while he favourably gave Africa resources, Africans must work against many odds.

I heard someone joke that God must have been in a very good mood when he created Africa. As someone who has spent some time on other continents, I attest. From a strategic symmetrical position about the equator, Africa boasts the best weather all year round and is favourably positioned to tap into the most fundamental source of energy globally: the sun. When we talk of the basic factor of production—fertile lands—Africa holds 65 per cent of the global arable land. Talk of rivers, seas, lakes, captivating sceneries, diverse exciting wildlife—it is all here. This is in addition to the tremendous, seemingly infinite mineral wealth in the continent. The preceding pages have attempted to inform you of your responsibility of being a good steward of the abundant resources on the continent. You are responsible for erasing the shame that is the paradox of desperate lack amidst plenty that has now become almost synonymous with Africa. Innovative volunteerism is the formula. Let's apply it together.

A quote that calls us all to reflection reads, "A people without the knowledge of their history, origin, and culture is like a tree without roots." Simply implied in this quote is the understanding that we cannot progress without knowing where we have come from and without a comprehension of the stakes.

In my discussions with pan-Africanists from diverse disciplines, nationalities, and creeds across the continent and globe, I have come to implicitly conclude that the open secret to development that our founding fathers failed to consider is that winning over self-determined rule was just but the tip of the iceberg. As leaders of countries on a continent that seemed by all accounts to be capable of self-sufficiency, hence the envy of many, they would have to overcome extreme odds to make anything significant out of their much-deserved independence and benefit from what their countries and continent offer.

They could not do it alone but needed partners. And the most fundamental and critically important partner to this end was their citizenry.

To make these partnerships count, they had to prioritise investing in developing the ordinary people in their countries and assign them to be guardians and custodians of development across the continent.

This is the significant blunder of our founding fathers. They looked to develop material resources without prioritising the people. Innovative volunteerism now takes this critical task a notch higher. As scripture underscores, iron sharpens iron. Innovative volunteerism bestows upon each of us an opportunity and duty to use what we have in our areas of enterprise as premium for targeted alliance building. The aim is enhancing the capacity of our fellow brothers and sisters and neighbours so we can jointly engage in the catalytic areas to collectively build wealth for our countries and continent. The priority and tangible areas of engagement are the EBAFOSA pillars, as discussed in chapter 7.

So I am calling you to usher yourself to action—not looking at your circumstances but at the glaring opportunities that our collective action in exercise of innovative volunteerism presents.

Why? Because by crawling, a child learns to stand. That having witnessed the effectiveness of innovative volunteerism, as shared in chapter 7, and the glaring opportunities to tap into it, as shared in chapters 5 and 6, we resolve to act with determination, to learn from past experiences and refine our actions to apply innovative volunteerism more effectively.

Millions of your colleagues across the globe are reading through this text and making a determined decision to act. Don't be a spectator when you have been called to leadership on behalf of a continent, on behalf of yourself and those yet to be born, on behalf of a planet. You are not acting in isolation but leveraging on team spirit, the power of togetherness. Your effort is magnified to mean much more, more than you may ever imagine. This means not only expanding your skills and enterprise but a broader agenda that concerns the well-being of a billion people—our African people. The growth of our economies is firmly founded on industrialisation and wealth creation that works with nature to safeguard our planet, our heritage from the creator. As a result, our people can all put food on their tables, can have money for medical care and educating their children, can have more than enough to spend and save. This is it, my friends. And I use the term *friend* to mean anyone, including those I will never meet, regardless of nationality, race, creed, gender—who is convicted of the urgent imperative of innovative volunteerism and is ready to act in its fruition. Innovative volunteerism is the only bond of friendship and comradery we can meaningfully share.

Innovative volunteerism, my comrades, is how we can reshape the development of the African continent. This is how we can change our mindset, shift the paradigm to truly drive transformational change for ourselves and those yet to be born. Yes, it is possible, and our collective efforts will make it so.

And for us to do it so, we must learn from the hard-working and selfless spirit of ants, which demonstrate the true spirit of innovative volunteerism. And for us to do it so, we must apply to ourselves, the very lessons by which we can selflessly spur ourselves to action and unlock what is worth more than what finances can achieve. We cannot therefore recoil from collectivism to individualism.

Let us use what we have through the spirit of innovative volunteerism to galvanise actions that can win the future for ourselves and for those yet to be born. We must therefore usher ourselves to action.

1. If there is a mother who cannot feed her newborn child with the proper food to live beyond the age of five, that should matter to us. If it matters to us, we must usher ourselves to action.
2. If there is an African who goes to bed with a stomach aching from hunger, that should matter to us. If it matters to us, we must usher ourselves to action.
3. If there is an unemployed youth whose skill could be tapped to drive the agro value chain for jobs and food security, that should matter to us. If it matters to us, we must usher ourselves to action

By ushering ourselves to action through the spirit of innovative volunteerism, we can ensure that Africa will never again experience the fear of want or need. The future is in our hands; we have everything it takes to shape it for the collective benefit of all. Let's seize this moment.

ABOUT THE AUTHOR

Driven by Africa's disproportionate affliction by climate change, Dr Richard Munang spearheaded a paradigm shift to building resilience, breaking out like wildfire across the continent. This paradigm is founded on four transformational premises. Foundationally is leveraging on climate action as an accelerator of socio-economic development as opposed to a silo environmental obligation as classically pitched. Secondly to drive this foundation, the approach is targeting productivity maximisation in catalytic sectors. These are sectors that are economically inclusive and capable of meeting leading socio-economic development priorities—especially food security, creation of income and job opportunities, and macro-economic expansion—simultaneously with meeting country climate objectives. Third is inclusivity, where all sectors of the economy and actors (state and nonstate; institutional and individual) are engaged for economywide synergy in maximising the catalytic sectors. Fourth is leveraging on people and their skills, talents, networks, and ongoing initiatives as the sovereign capital and source of value to drive this strategic paradigm. Winner of an Africa Environmental Hero Award, Dr Munang holds a PhD in environmental change and policy from Nottingham University and was awarded an Executive Education Certificate having undertaken Harvard University's Executive Programme in Climate Change and Energy Policy Making for the Long Term at the Kennedy School. An avid social media user for knowledge sharing, Dr Munang tweets regularly; his handle is @RichardMunang.